Motion

Motion involves the change of position of a body with time. The study of the motion of a body without involving the force which causes motion is called **kinematics**. On the other hand, the study of the motion of a body with regards to the cause of motion is called **dynamics**.

In this chapter, we are going to study the kinematics of bodies in straight line

1 Types of Motion

Plan 1

1

The following are the types of motion a body can undergo:
- (a) Random motion
- (b) Translational motion
- (c) Rotational motion
- (d) Oscillatory motion
- (e) Rectilinear motion

We will explain each of them briefly.

Random motion

2

When a body moves in a zig-zag manner with no specific pattern, the body is said to undergo **Random motion**. Examples are motion of molecules of gases, butterflies, moths and smoke particles.

You could easily pick the correct answer for this JAMB question

| 3 |

The motion of smoke particles from a chimney is typical of
(A) random motion (B) circular motion (C) rotational motion (D) oscillatory motion.

Answer: If you chose random motion, then you are right!

Translational motion

| 4 |

This is the motion of a body in a particular direction from one point to another. An example is a car moving from one town to another in a particular direction.

Rotational motion

| 5 |

When a body moves about an axis through itself, the body is said to perform rotational motion. Examples are: (i) the wheels of a moving car, (ii) the rotating blades of an electric fan, and (iii) the rotation of the earth about its axis.

Oscillatory motion

| 6 |

When a body moves to and fro about a fixed point, it performs an oscillatory motion. Examples are: the simple pendulum, a plucked guitar string, vibration of molecules in a solid.

7

This is motion in a straight line. An example is light ray travelling from one point to another along a straight path.

2 Parameters used in describing Motion in a straight line

Plan 8

8

We can describe the motion of objects in a straight line using the following terms/parameters:

(i) Distance and Displacement
(ii) Speed and Velocity
(iii)   Acceleration

Displacement/Distance

9

Displacement is distance in a specified direction; it is therefore a vector quantity.
The major difference between distance and displacement is that distance does not require specifying the direction of motion (it is a scalar) while displacement requires specifying the direction (it is a vector).

They both have the same unit, which is **meter (m).** We illustrate how displacement differs from distance below.

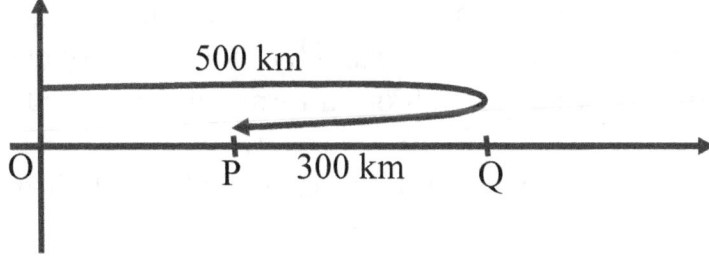

Figure 1

As shown in the diagram above, when a body moves from point O to point Q and then to point P in the direction indicated by the arrow, it covers a total distance of 800km while its displacement form the origin is only 200km in the East direction.

Speed/Velocity

| 10 |

Speed is the rate of change of distance, that is:

$$\text{Speed} = \frac{\text{change in distance}}{\text{time}}$$

Velocity is the rate of change of displacement, that is:

$$\text{Velocity} = \frac{\text{change in displacement}}{\text{time}}$$

Velocity is therefore speed in a specified direction, it is a vector quantity.

The difference between speed and velocity is similar to the difference between distance and displacement; **speed is a scalar while velocity is a vector.**
They both have the same unit, which is meter per second $(\textbf{ms}^{-1})$.

The concept of uniform velocity/speed

| 11 |

If the change in displacement/distance with time is constant, then the velocity/speed is said to be uniform. More candidly, if a car is not changing its velocity/speed, it is said to have a constant or uniform velocity/speed.

If we say for instance that a car has a uniform speed of 50km/h, it means that the car covers a distance of 50km in every one hour, and that the car never moved slower or faster, but maintained this same speed.

Uniform acceleration

The acceleration is said to be uniform if the change in velocity with time is kept constant, that is, if the acceleration is not changing.

As an illustration, suppose we say that a car is moving with a uniform acceleration of $30km/h^2$, it means that the car is increasing its velocity by 30km/h in every one hour, and that this increment rate never changes.

Retardation

Retardation (also called deceleration) is negative acceleration. It has the same unit as acceleration. Usually retardation is quoted as acceleration but in negative magnitudes.

For instance, if we say that the acceleration of a body is $-6ms^{-2}$, it means that it is decelerating at $6ms^{-2}$. The velocity of this body is decreasing by $6ms^{-1}$ in every one second.

3 Distance/Displacement – time graphs

Plan 14

The distance/displacement-time graph of an object is a graph illustrating how the distance/displacement of that object changes with time. It is a graph with distance/displacement on the vertical axis, and time on the horizontal axis.

If a body is not moving, its displacement/distance will not change, and so the displacement/distance-time graph will be horizontal as shown in figure 2(a).

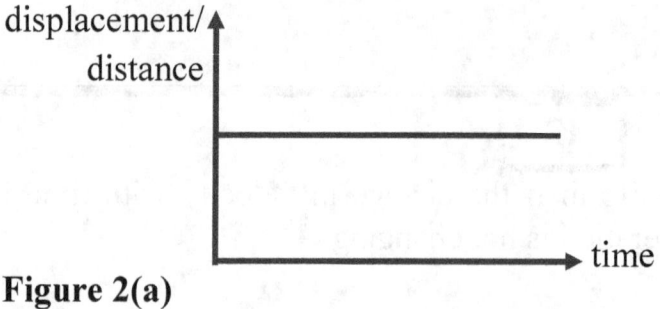

Figure 2(a)

If the body is moving with a uniform speed, the graph will be linear and have some slope as shown in figure 2(b), but if however the body is moving with a non-uniform speed, then the graph will not be linear as shown in figure 2(c).

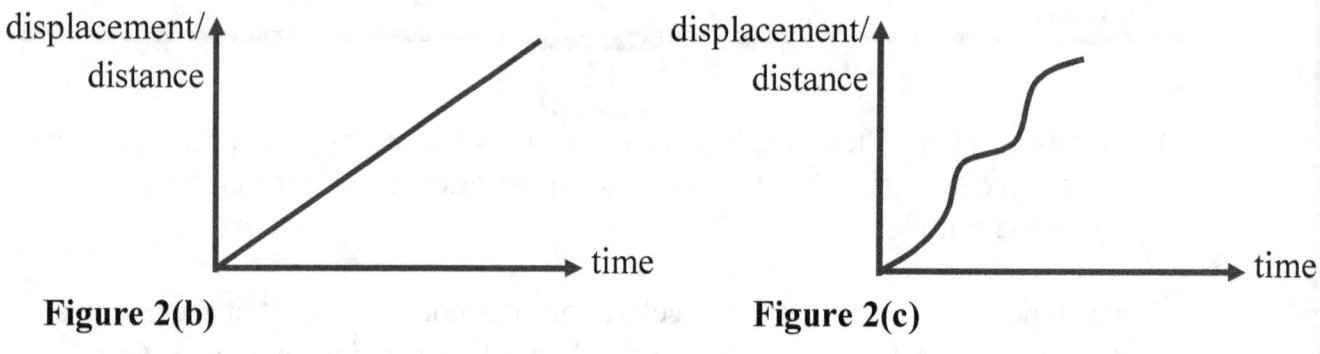

Figure 2(b)   **Figure 2(c)**

Important!

15

The slope of a displacement/distance-time graph at any point in the graph represents the velocity/speed at that point.

Yes, if the graph is linear, then its slope is a constant value, this represents the constant speed of the body. If however the graph is non-linear, that means the speed is not uniform, and so we can find the speed at any point by measuring the slope of the tangent to the curve at that point.

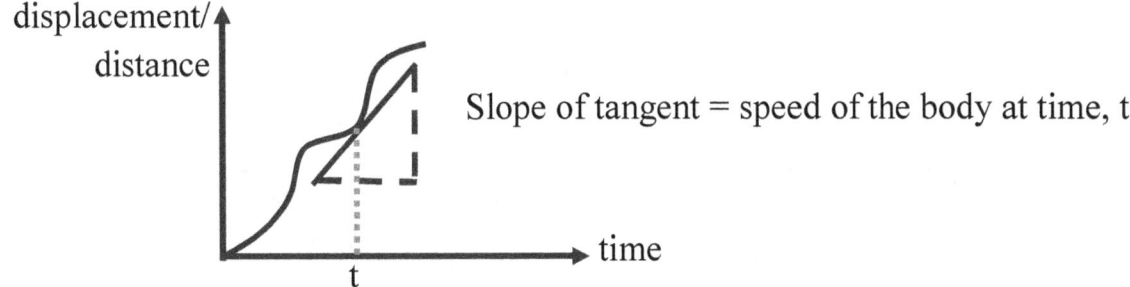

Slope of tangent = speed of the body at time, t

Figure 3

4 Velocity - time graphs

Plan 16

| 16 |

The velocity-time (v-t) graph of an object is a graph illustrating how the velocity of the object changes with time. It is a graph of velocity on the vertical axis and time on the horizontal axis.

If the object is moving with a constant speed, the v-t graph will be horizontal as shown in figure 4(a). If it is moving with a uniform acceleration, then we have a linear graph as in figure 4(b), and if the acceleration is non-uniform, we will have a non-linear graph, figure 4(c) is an example. If the object is decelerating uniformly, we get a linear graph with negative slope as in figure 4(d).

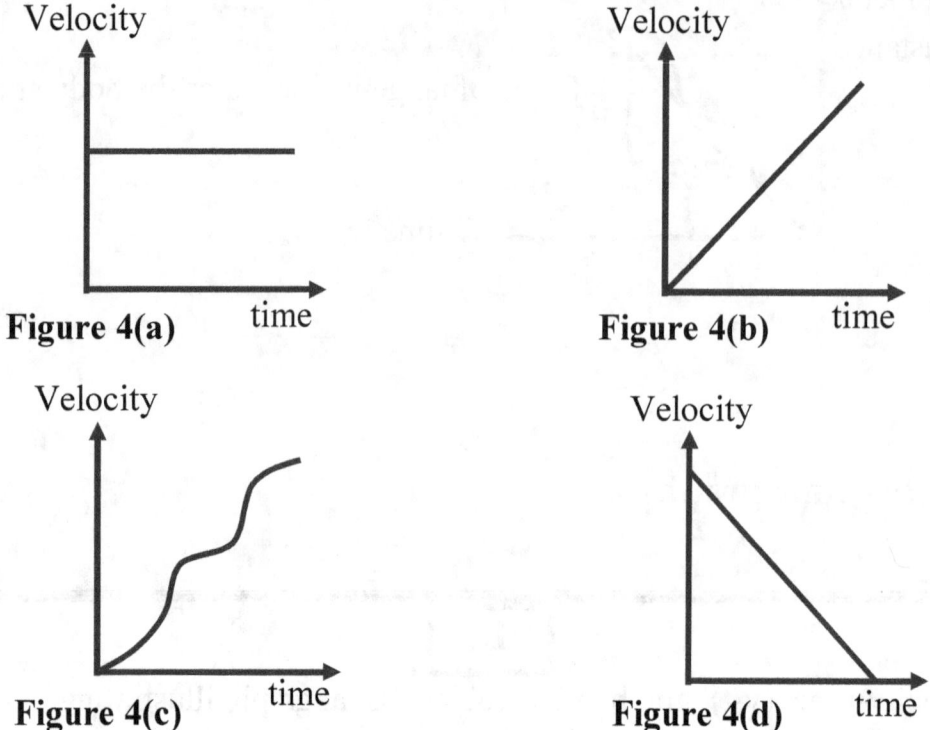

Figure 4(a)   Velocity / time

Figure 4(b)   Velocity / time

Figure 4(c)   Velocity / time

Figure 4(d)   Velocity / time

I'm pretty sure you know what the graph will look like if the deceleration is non- uniform. If you guessed something like the diagram in plan 18, then you are right!

Plan 17

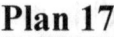

17

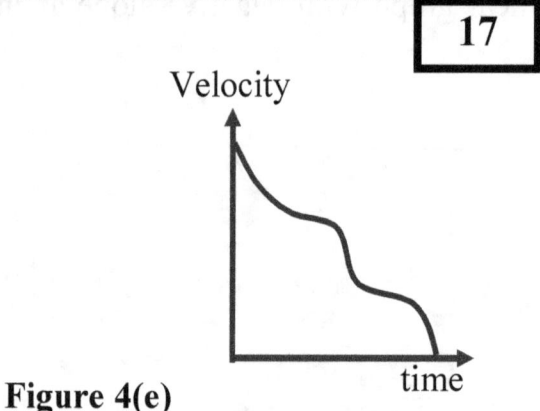

Figure 4(e)

Important for v-t graphs:

1. The slope of a v-t graph at any point in the graph represents the **acceleration** at that point.

2. The area under the v-t graph represents the **distance** covered within that period

We'll take some examples to illustrate all this later!

This JAMB question is a walk over!

$$\boxed{18}$$

The velocity – time graph of a body moving in a straight line and decelerating uniformly to rest is represented by

(A) (B) (C) (D)

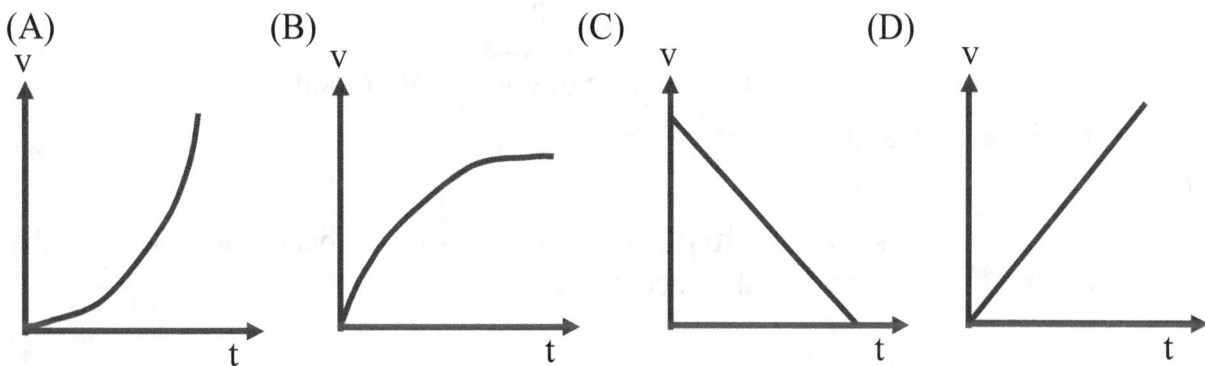

If you chose option C, then you are right!

5 Equations of a uniformly accelerated motion

Let's go!

The motion of particle that has a constant acceleration is referred to as uniformly accelerated motion. This is the type of motion we want to describe in equations.

We will consider an object moving with uniform acceleration a, such that the velocity changes from an initial value u, to a final value v, in a time t, and it covers a distance s, within the period.

Plan 20

By definition, acceleration is the rate of change of velocity.

That is, acceleration $= \dfrac{change\ in\ velocity}{time}$

If u and v represent the initial and final velocities respectively, then the change in velocity is v-u, and so the acceleration a is

$$a = \frac{v-u}{t}$$

Giving us equation (1) if we make v subject of the formula

$$v = u + at \tag{1}$$

Next, the total distance covered is defined as the average velocity multiplied by the total time taken, that is

total distance (s) = average velocity × time (t)

our average velocity is the sum of the two velocities divided by 2 (that is $\frac{u+v}{2}$)

And so total distance (s) $= \dfrac{u+v}{2} \times$ time (t)

Giving us equation (2)

$$s = \left(\frac{u+v}{2}\right)t \qquad\qquad (2)$$

Alternatively, we could also obtain equations 1 and 2 graphically

21

Here we still consider our object accelerating uniformly from an initial velocity, u to a final velocity, v in a time, t. Its velocity-time graph will be as shown in figure 5.

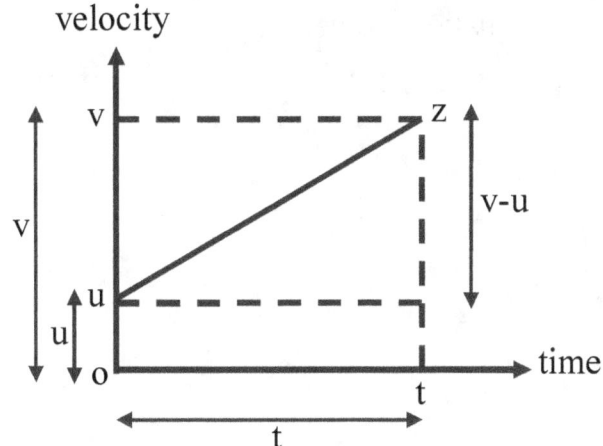

Figure 5

And as we explained in plan 18, the acceleration of this object will be given by the slope of line uz, that is:

a = slope of line uz = $\frac{v-u}{t}$

And if we make v subject of the formula, we get equation 1

v = u + at   (1)

Next, we also explained in plan 18 that the distance covered will be given by the area under the graph, that is the area of trapezium ouzt:

s = area of trapezium ouzt = ½(u + v)×t
which is the same as equation 2

$$s = \left(\frac{u+v}{2}\right)t \qquad\qquad (2)$$

Let's go on! 2 more equations!

22

Eliminating t from equations 1 and 2 will give us the next equation:
$$v^2 = u^2 + 2as \tag{3}$$

How? First we make t subject of the formula in equations 1 and 2 to get $t = \frac{v-u}{a}$ and $t = \frac{2s}{u+v}$ respectively.

Next we equate the two equations to get:
$$\frac{v-u}{a} = \frac{2s}{u+v}$$

cross multiplying gives:
$$(v-u)(u+v) = 2as$$
$$vu + v^2 - u^2 - vu = 2as$$

which gives equation 3:
$$v^2 = u^2 + 2as \tag{3}$$

Then the last one,...

23

We eliminate v from equations 1 and 2 to get the last equation:
$$s = ut + \tfrac{1}{2}at^2 \tag{4}$$

How?
v is already subject of the formula in equation 1, if we make v subject of the formula in equation 2 we get $v = \frac{2s}{t} - u$

Equating this to equation 1 gives:
$$\frac{2s}{t} - u = u + at$$

$$\frac{2s}{t} = 2u + at$$

Multiplying through by t, we get
$$2s = 2ut + at^2$$

And dividing through by 2 gives
$$s = ut + \frac{1}{2}at^2 \qquad (4)$$

In summary,...

24

Equations (1) to (4) are called equations of motion.

Equation (2) is often removed from the list of equations of uniformly accelerated motion because it does not contain the acceleration term, a.

We list all the four equations again to make it easy for you to memorize:

$$v = u + at \qquad (1)$$
$$s = \left(\frac{u+v}{2}\right)t \qquad (2)$$
$$v^2 = u^2 + 2as \qquad (3)$$
$$s = ut + \frac{1}{2}at^2 \qquad (4)$$

They are useful in solving problems associated with uniformly accelerated motion.

Let's begin with this JAMB question

25

A body starts from rest and moves with uniform acceleration of 6ms^{-2}. What distance does it cover in the third second?

(A) 15m   (B) 18m   (C) 27m   (D) 30m

Let's do it!

26

We are told that the body starts from rest, and so it's initial velocity, u=0, and we are given that the acceleration is 6ms^{-2}.

Now we need to get the distance it covers in the third second, which is the distance covered at time, t = 3s minus distance covered at time, t = 2s.

We need to use the equation: $s = ut \pm \frac{1}{2} at^2$ since we are looking for the distance, s, and we know the other parameters.

For t = 3 seconds, we get
$s = 0 + \frac{1}{2} \times 6 \times (3)^2 = 6 \times 9/2 = 27m$

And for t = 2 seconds:
$s = 0 + \frac{1}{2} \times 6 \times (2)^2 = 12m$

And so the distance covered in the third second is equal to 27m – 12m = 15m

You got it? Then move on!

Plan 27

27

A car moves from rest with an acceleration of 5m/s^2. Find its velocity when it has moved a distance of 50m.

Solution:
In this question, we are looking for the final velocity, v.
Again, we are told that the car moves from rest, so its initial velocity, u=0. We are also given the acceleration, a=5m/s^2, and the distance, s=50m

An appropriate formula will be $v^2 = u^2 + 2as$
Therefore $v = \sqrt{u^2 + 2as}, \quad = \sqrt{0^2 + 2 \times 5 \times 50} \quad = \sqrt{500} \quad = 10\sqrt{5} \, m/s$

28

A car has a uniform velocity of 60 km/h. How far does it travel in 1 minute?

Solution:

In this question, the car has a uniform velocity which means zero or no acceleration, so we can directly solve using the formula that:

$$velocity = \frac{distance}{time}$$

But before that, we'll like to convert the units in the question to S. I. units so as to have conformity; the unit of the velocity has to be m/s and the unit of the time should be s, so that the unit of the resulting distance will be m.

velocity = 60 km/h = $\frac{60 \times 1000}{60 \times 60}$ = 16.7 m/s   [Notice that we multiply by 1000

to convert the km to m, and divide by (60 × 60) to convert the per hour to per seconds]

Also the time = 1 minute = 60 seconds

And using $velocity = \frac{distance}{time}$, we get:

distance = velocity × time   = 16.7 × 60   = 1002m ≈ 1 km

NOTE that an alternatively fast way to deal with this problem will have been to say that since the speed of the car is 60 km/h, it means that the car covers 60 km in every hour OR 60 km in every 60 minutes, and so it covers 1 km in every 1 minute. And that answers the question: the distance it travels in 1 minute is 1 km.

If that is OK, we can fire on!

Graphical approach to uniformly accelerated motion

29

Problems on uniformly accelerated motion can be solved graphically by use of a velocity–time (or speed–time) graph as we shall illustrate below.

We will use a couple of problems to show how this is done!

Example 1

30

A body starts from rest and accelerates uniformly in a straight line until it reaches a velocity of 40ms^{-1} in 8 seconds. It then travels uniformly (constant speed) for 12 seconds after which it is uniformly retarded and brought to rest in 5 seconds calculate (i) the acceleration, (ii) the retardation, and (iii) the total distance covered in the entire motion.

Here we go!

31

First, we need to represent the information contained in this question graphically. And as the first question of this kind, we shall do so in very bit-by-bit steps for easy understanding. OK, let's do it!

First, the body started from rest and accelerated uniformly until it reaches a velocity of 40ms^{-1} in 8 seconds:

Next the body travels with a constant speed for 12 seconds (making a total of 8+12 seconds at this point):

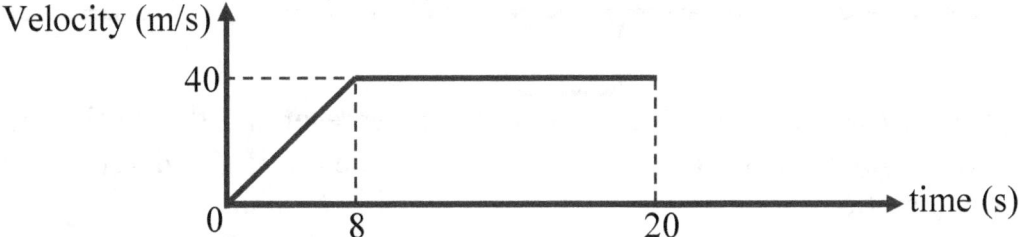

And finally, it is uniformly retarded and brought to rest in 5 seconds (making the total time to be 8+12+5):

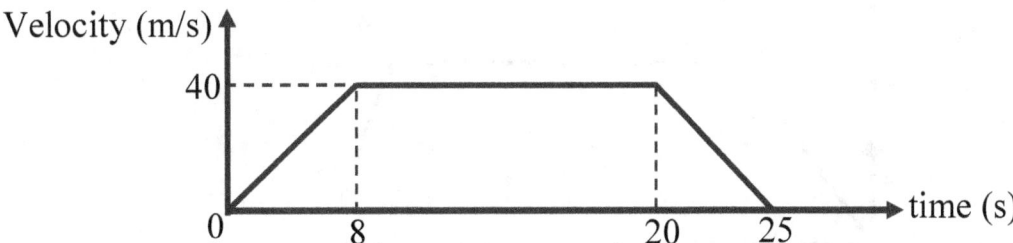

That is it! And for convenience, we shall label the diagram as shown below.

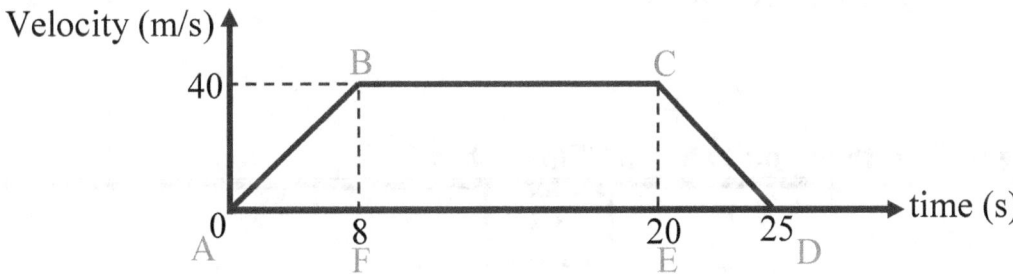

Figure 6

Now, the answers:

  (i) The acceleration is the slope of line AB = 40/8 = 5m/s^2

  (ii) The retardation is the slope of line CD = 40/(25-20) = 40/5 = 8m/s^2

  (iii) The total distance covered is equal to the Area of the Trapezium ABCD
      = ½ (BC + AD) × BF
      = ½ (12 + 25) × 40
      = 740m

32

The diagram below shows a velocity–time graph representing the motion of a car. Find the total distance covered during the acceleration and retardation periods of the motion.

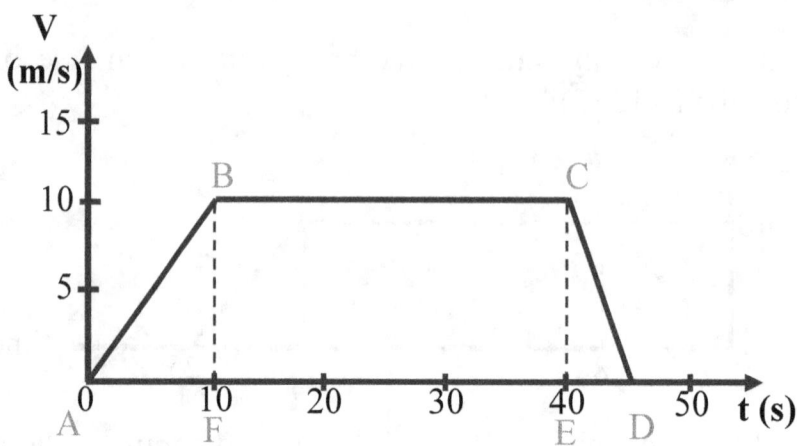

(A) 75m   (B) 150m   (C) 300m   (D) 373m

If you got 75m, then you're right! This is how!

33

Distance travelled during the acceleration period is given by area of the triangle ABF
= ½ AF × BF   = ½ × 10 × 10 = 50m

And the distance travelled during the retardation period is given by area of the triangle CDE
= ½ ED × CE   = ½ × 5 × 10   = 25m

Therefore the total distance travelled in these 2 periods is (50 + 25)m = 75m

34

Before we illustrate how derivatives and integrals are applied in motion, we want to use this plan to introduce you to the core and important concepts.

The subject of Calculus (broadly classified as Derivatives and Integrals) is usually taught extensively during the early years of Science/Engineering studies at higher institutions, so if you've never heard about this before, don't break your head! Just relax and grab what we have to say here, it is all you need at this stage!

The derivative of a function (also called the differentiation of that function) usually tells us the rate at which that function is changing. The Integral (also called the integration) of a function is the reverse process, and so occasionally referred to as anti-differentiation.

Now this is how it all comes into motion: we define acceleration as the rate of change of velocity, this means that acceleration is the derivative of velocity. And we define velocity as the rate of change of displacement, which means that velocity is the derivative of displacement.

Let's make it easy to grab here!

35

These 3 parameters: displacement, velocity, and acceleration can be ranked in that order; displacement → velocity → acceleration
So that the differentiation of displacement gives velocity, and the differentiation of velocity gives acceleration.

The reverse is also true for integration: the integration of acceleration gives velocity, and the integration of velocity gives displacement.

Notation

The derivative of velocity, v, with respect to time, t, is written as $\frac{dv}{dt}$, and the integral of v with respect to t is written as $\int v\ dt$.

We can therefore write mathematically that:

acceleration a $= \frac{dv}{dt}$, and velocity v $= \frac{ds}{dt}$ where s is displacement.

And on the other hand:

velocity v $= \int a\ dt$, and displacement s $= \int v\ dt$

Quick clues to Differentiation

In general, if y=ax^n, then the derivative of y with respect to x is:

$$\frac{dy}{dx} = nax^{n-1}$$

Specifically, if we are given y $= 3x^4 + 2x^3 - 5x^2 + 4x - 7$, then the derivative of y with respect to x will be:

$\frac{dy}{dx} = 12x^3 + 6x^2 - 10x + 4$

And that is it! Simply multiply the coefficient of x with the power of x, and then decrease the power by 1.

Quickly differentiate the following with respect to x.

1. $12x^5 - 20x^3 - 15x^2 + 40x + 200$

2. $30x^4 + 12x^3 - 50x^2 + 27x - 1$

3. $3x^{11} + 2x^9 - 5x^6 + 1$

4. $25x^3 - 2x^2 + x - 12$

5. $3 + 2x^4 - 2x - 5x^2 + 4121x^3$

If you got the following answers, then you are good to go:

1. $60x^4 - 60x^2 - 30x + 40$

2. $120x^3 + 36x^2 - 100x + 27$

3. $33x^{10} + 18x^8 - 30x^5$

4. $75x^2 - 4x + 1$

5. $8x^3 - 2 - 10x + 12363x^2$

Now we are set to take some questions!

Let's start with this JAMB question

39

The velocity v of a particle in a time t is given by the equation: $v = 10 + 2t^2$. Find the instantaneous acceleration after 5 seconds.

(A) $60ms^{-2}$ (B) $20ms^{-2}$ (C) $15ms^{-2}$ (D) $10ms^{-2}$

Solution

40

First, we are given the velocity, and we are looking for the acceleration, so we have to differentiate the velocity to get the acceleration. Doing this gives:

$v = 10 + 2t^2$

Therefore, acceleration, $a = \frac{dv}{dt} = 4t$

Then at time, t = 5 seconds, the acceleration will be

$a = 4t = 4 \times 5 = 20ms^{-2}$

If you got it, we shall proceed with the integration aspect, and later take more questions.

Quick clues to Integration!

41

In general, if $y = ax^n$, then the integral of y with respect to x is:

$$\int y\,dx = \frac{ax^{n+1}}{n+1}$$

Yes, just add 1 to the power, and divide by the new power!

Specifically, if we are given $y = 12x^3 + 6x^2 - 10x + 4$, then the integral of y with respect to x will be:

$\int y\, dx = \frac{12x^4}{4} + \frac{6x^3}{3} - \frac{10x^2}{2} + \frac{4x^1}{1} + C$, where C is a constant usually added to integration results.

In simplified form:

$\int y\, dx = 3x^4 + 2x^3 - 5x^2 + 4x + C$

And that's it!

JUST A NOTE: Observe that differentiating this result will bring us back to

$y = 12x^3 + 6x^2 - 10x + 4$, and that is why we said that integration is anti-differentiation; integration and differentiation are reverse processes of each other.

Now, see if you got that!

$\boxed{42}$

Integrate the following with respect to x:

1. $12x^5 - 20x^3 - 15x^2 + 40x + 200$

2. $30x^4 + 12x^3 - 50x^2 + 27x - 1$

3. $3x^{11} + 2x^9 - 5x^6 + 1$

4. $25x^3 - 2x^2 + x - 12$

5. $3 + 2x^4 - 2x - 5x^2 + 4121x^3$

Congratulations if you got the following answers! (If not, just take a close look at your solutions again to observe what you did wrong).

1. $2x^6 - 5x^4 - 5x^3 + 20x^2 + 200x + C$

2. $6x^5 + 3x^4 - \frac{50}{3}x^3 + \frac{27}{2}x^2 - x + C$

3. $\frac{1}{4}x^{12} + \frac{1}{5}x^{10} - \frac{5}{7}x^7 + x + C$

4. $\frac{25}{4}x^4 - \frac{2}{3}x^3 + \frac{1}{2}x^2 - 12x + C$

5. $3x + \frac{2}{5}x^5 - x^2 - \frac{5}{3}x^3 + \frac{4121}{4}x^4 + C$

Now, let's take a problem!

43

The velocity of an object is defined by $v = 3t^2 - 2t + 1$, where v is velocity in m/s, and t is time in seconds. Find

   (i) the distance it covers between the third and fourth seconds

   (ii) the distance it covers in the first 10 seconds

   (iii)   its acceleration after 10 seconds

Solution

44

First, we have known that to get distance from velocity we have to integrate, and to get acceleration from velocity we have to differentiate, so

distance $s = \int v\, dt = t^3 - t^2 + t + C$

and acceleration $a = \frac{dv}{dt} = 6t - 2$

(i) Now, distance covered after 3 seconds $= 3^3 - 3^2 + 3 + C$

$= 27 - 9 + 3 + C = 21 + C$

And distance covered after 4 seconds $= 4^3 - 4^2 + 4 + C$

$= 64 - 16 + 4 + C = 52 + C$

Therefore distance covered between the 3rd and the 4th seconds is:

$(52 + C) - (21 + C) = 52 - 21 + C - C = 31m$

NOTE that this problem can also be solved by method of definite integrals as follows:

Distance covered between the 3rd and the 4th seconds $= \int_3^4 v \, dt$

$= \int_3^4 (3t^2 - 2t + 1) \, dt = t^3 - t^2 + t \, \big|_3^4$

$= (4^3 - 4^2 + 4) - (3^3 - 3^2 + 3) = 52 - 21 = 31\text{m}.$

The constant of integration C is always omitted in definite integral; that's the only difference!

(ii) The distance covered in the first 10 seconds = distance covered between t=0 and t=10 seconds $= \int_0^{10} v \, dt$

$= \int_0^{10} (3t^2 - 2t + 1) \, dt = t^3 - t^2 + t \, \big|_0^{10}$

$= (10^3 - 10^2 + 10) - (0^3 - 0^2 + 0) = 910 - 0 = 910\text{m}.$

(iii) The acceleration after 10 seconds $= \frac{dv}{dt} \big|_{at\ t=10s} = 6t - 2 \big|_{at\ t=10s}$

$= 6(2) - 2 = 10\text{m/s}^2$

6 Motion under gravity

Plan 45

45

Neglecting air resistance, motion under gravity is a typical sample of uniformly accelerated motion with acceleration g (the acceleration due to gravity).

The vertical fall of all unsupported objects near the surface of the earth with constant acceleration due to gravity g (if air resistance is neglected), is referred to as free fall.

Positive and Negative acceleration

46

The acceleration due to gravity is positive for objects falling downwards because as objects fall, gravity increases their velocities, while it is negative for objects moving upwards because objects moving upwards are going against gravity and so gravity decreases their velocities.

So, we can re-write our equations (1), (3) and (4) by replacing a with ±g as follows:

$$v = u \pm gt \tag{1b}$$

$$v^2 = u^2 \pm 2gh \tag{3b}$$

$$h = ut \pm \tfrac{1}{2} gt^2 \tag{4b}$$

Just a few notes!

47

The following points have to be noted also in motion under gravity.

(i) If a body is simply dropped from a height, we put $u = 0$, because its motion starts from rest, and so zero initial velocity.

(ii) When it is projected vertically upwards, the velocity when it reaches its maximum height $v=0$, because at maximum height it stops moving just before it begins falling down.

(iii) When the body falls to the ground once again the height h above the ground is zero, $h = 0$.

Example

48

A body is dropped from rest at a height of 80m. How long does it take to reach the ground? Ignore air resistance ($g = 10ms^{-2}$).

Solution:

From equation 4b:

$h = ut + \frac{1}{2}gt^2$

Where h = 80m, $g = 10ms^{-2}$, u=0, and t =?

$80 = 0 + \frac{1}{2} \times 10 \times t^2$

$80 = 5t^2$

$t^2 = 80/5 = 16$

Therefore, $t = \sqrt{16} = 4s$

Exercises

49

1. An aeroplane lands on a runway at a speed of 180km/hr and is brought to a stop uniformly in 30 seconds. What distance does it cover on the runway before coming to rest?

(A) 360m   (B) 540m   (C) 750m   (D) 957m

(JAMB)

2. When a ball rolls on a smooth level ground, the motion of its centre is

(A) translational   (B) oscillatory   (C) random   (D) rotational (JAMB)

3. Which quantity X is calculated using this equation?

$$X = \frac{change\ in\ velocity}{time\ taken}$$

(A) acceleration (B) average velocity (C) distance travelled (D) speed
(Cambridge)

4.

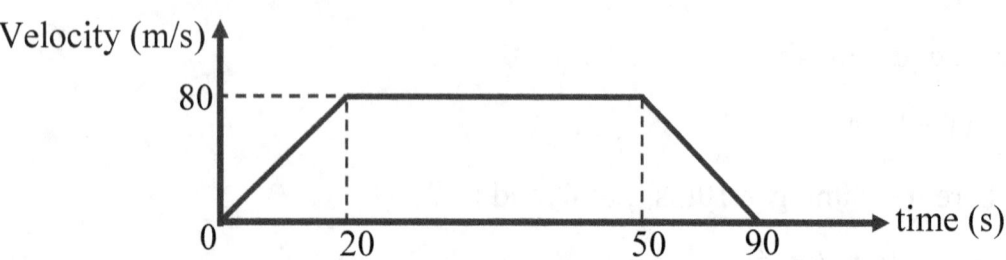

The diagram above shows the velocity – time graph of a vehicle. Its acceleration and retardation respectively are

(A) 8.0ms^{-2}, 4.0ms^{-2}

(B) 4.0ms^{-2}, 8.0ms^{-2}

(C) 4.0ms^{-2}, 2.0ms^{-2}

(D) 2.0ms^{-2}, 4.0ms^{-2}

(JAMB)

5. In free fall, a body of mass 1kg drops from a height of 125m form rest in 5s. How long will it take another body of mass 2kg to fall from rest from the same height?

(A) 5s   (B) 10s   (C) 12s   (D) 15s

(JAMB)

6. A motor vehicle is brought to rest from a speed of 15ms^{-1} in 20 seconds. Calculate the retardation.

(A) 0.75ms^{-2} (B) 1.3ms^{-2} (C) 5.0ms^{-2} (D) 7.5ms^{-2}

(JAMB)

7. The distance m travelled by a particle in time t is described by the equation m = 10 + 12 t^2. Find the average speed of the particle between the time interval t = 2s and t = 5s

(A) 60ms^{-1} (B) 72ms^{-1} (C) 84ms^{-1} (D) 108ms^{-1}

(JAMB)

8. The graph shows the movement of a car over a period of 50 s.

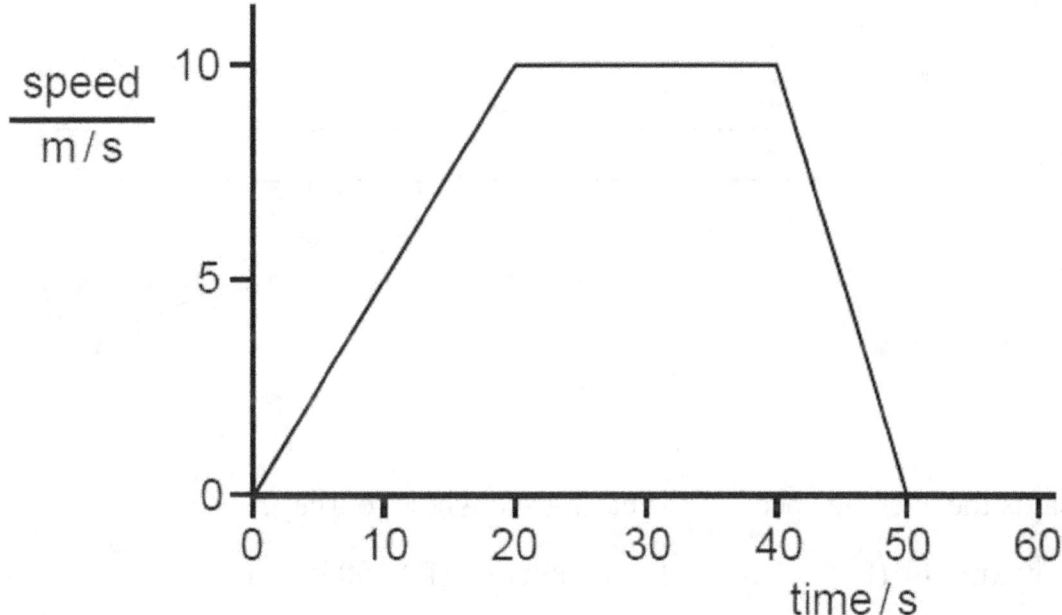

What was the average distance travelled by the car while its speed was increasing?

(A) 10 m   (B) 20 m   (C) 100 m   (D) 200 m

(Cambridge)

9. Which graph represents the motion of a body falling vertically that reaches a terminal velocity?

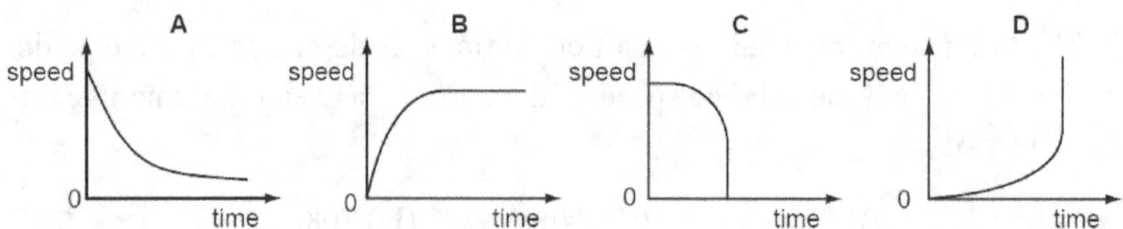

(Cambridge)

10. A car takes 1 hour to travel 100 km along a main road and then ½ hour to travel 20 km along a side road.

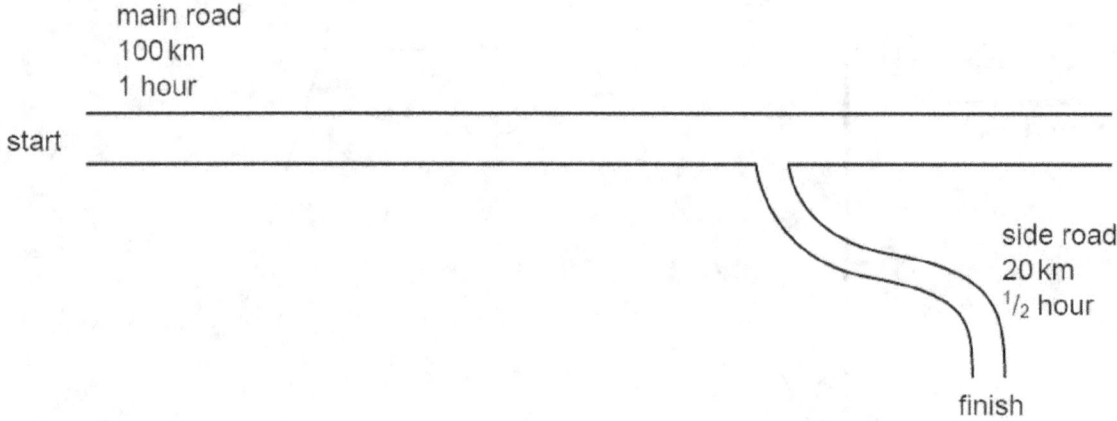

What is the average speed of the car for the whole journey?

(A) 60 km / h   (B) 70 km / h   (C) 80 km / h   (D) 100 km / h

(Cambridge)

50

1. C

2. A

3. A

4. C

5. A

6. A

7. C

8. C

9. B

10. C

www.ingramcontent.com/pod-product-compliance
Lightning Source LLC
Chambersburg PA
CBHW081246170526
45165CB00009B/3216